Scientific Papers

The Scientific Papers

Nathan Coppedge

PREFACE

This hodgepodge collection of experiments and eclectic observations was formed over the period 2005 - 2014, during which I formed at least two major discoveries: (1) That over-unity is possible, vis. An experiment with a trough-leverage apparatus in which a marble is raised upwards from a position of rest and then triggers the counterweight on its own, returning all parts to their initial altitudes, in spite of the motion which occurred, and (2) A remarkable subtle arrangement of two pieces of cardboard allowing a marble to run a course that is not downwards, but upwards, of its own volition.

There were other accomplishments as well during this remarkable time in my life. See for yourself!

Experiments on the fringe of physics, daring to question the rudiments of reality, as we know it conventionally.

RED LETTER DAYS IN THE PURSUIT OF
PERPETUAL MOTION

JULY 3, 2014: Day I discovered that objects can
roll upwards, using a "Master Angle." Based on
an earlier experiment with the Ersatz Coquette (June 29, 2014).

NOV 10, 2013: My First Successful Over-Unity
Experiment (AcademicRoom.Com) or view it
at my blog instead If a device can go up and
then down from a position of rest, and all parts
return to their original positions, then there is
theoretical over-unity. This device meets the
criteria. The diagram shows how to turn the
over-unity device into a perpetual motion machine.

APRIL 26TH, 2009: Day I Invented 50 Devices
These devices contributed significantly to my
working tools, but so far as I know none of
them have joined my online collection. I cite
many flaws.

APRIL 16TH: Day in which I have invented
Five Devices more than once, first in 2007 One
of these was the Coquette.

APRIL 2ND, 2007: Day I found evidence that
the Tilt Motor could work. With near horizontal slope, leverage can extend slope. Confirmed once, denied twice. Confirmation was
with a level.

FEBRUARY 4TH, 2007 Day I seemed to prove an unbalanced wheel, the Principled Asymmetry. The device seems to rotate more easily in one direction than another.

OCTOBER 30TH, 2006: Day I invented the Tilt Motor, and time-traveled back to the morning of the same day. The Tilt Motor is the ingenious concept of a horizontal wheel, operated by levers.

CIRCA MARCH 2005: A visit to Mystic Aquarium resulted in an encounter with a 'Marble Rally' game, inspiring thoughts about leverage with supporting slope. However, the actual device concept was not formulated until years later.

MOTIVE MASS MACHINE EXPERIMENT 1

Basic proof was provided that a free falling weight can move an equal weight upwards, when the second weight is partially supported and rolls, e.g. because a rolling weight provides less resistance than a fixed weight, and in this case the fixed weights had an equi-librous effect.

An experiment in which a seesaw apparatus similar to a large motive-mass type seesaw was used to approxi-mate the proportions.

The low-friction pulley that was used to approximate ideal conditions.

MOTIVE MASS MACHINE EXPERIMENT 2

Here a smaller-scale tilting structure is used, this time testing whether the weight of a cart can be lifted by applying equal weight to one end of a balanced balance.

The cart used in the experiment

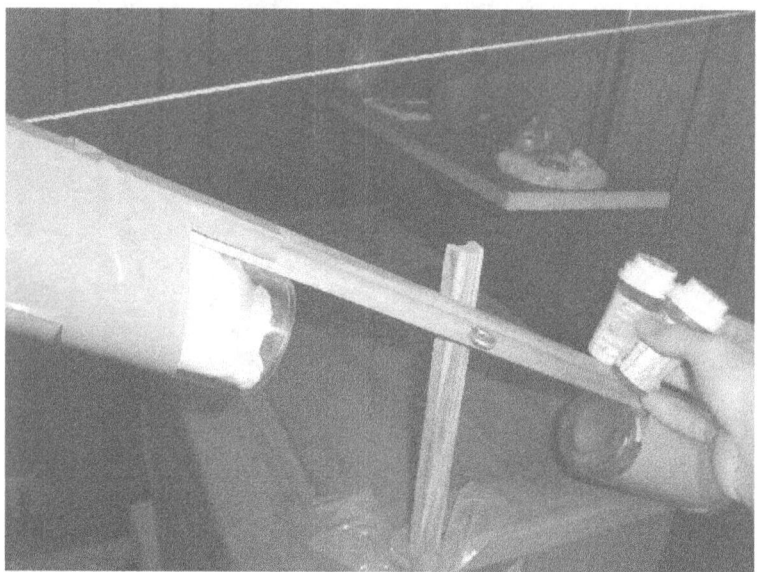

The equal weight was applied, and caused upwards motion. In theory the design is equally applicable at lesser or greater heights, due to the extensivity provided by the pulley system.

PRINCIPLED ASYMMETRY

The purpose of this device is to provide exponential momentum on inputted energy.

A. The heavier end adequately moves the slightly lighter double-armed end.
B. The double armed end approaches azimuth.
C. The heavy end is now supported briefly only by the axle.
D. The slightly lighter ends are thrown forward, without needing to counterbalance.
E. The heavier end is thrown upwards by the excess momentum.
F. If energy is inputted, the process gains momentum very quickly.

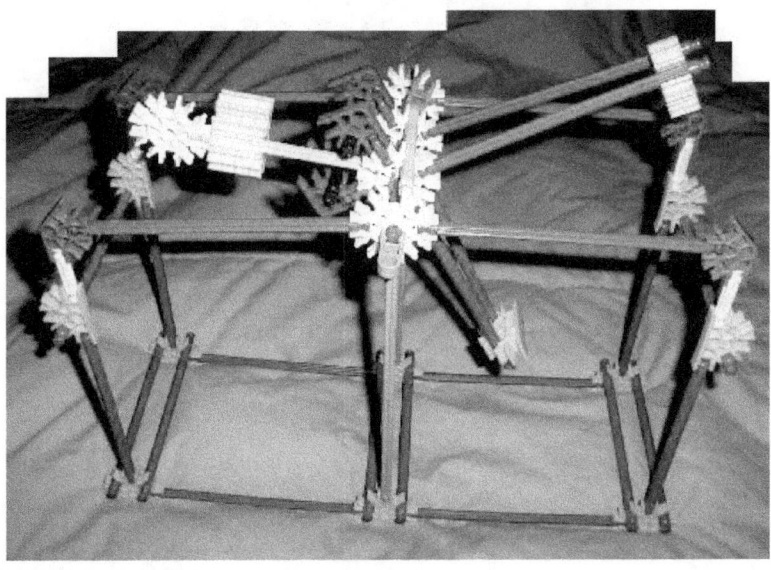

FRONT VIEW

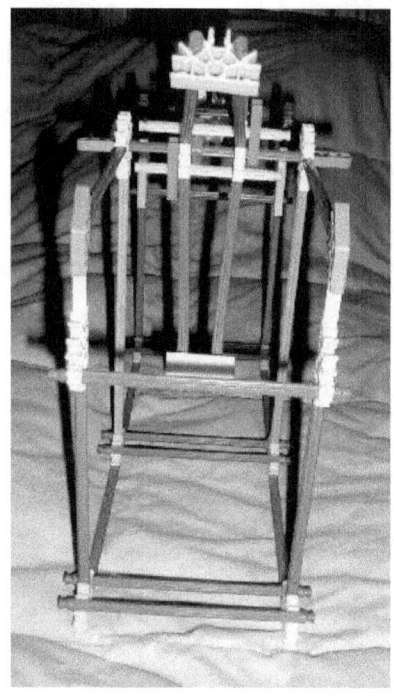

SIDEVIEW

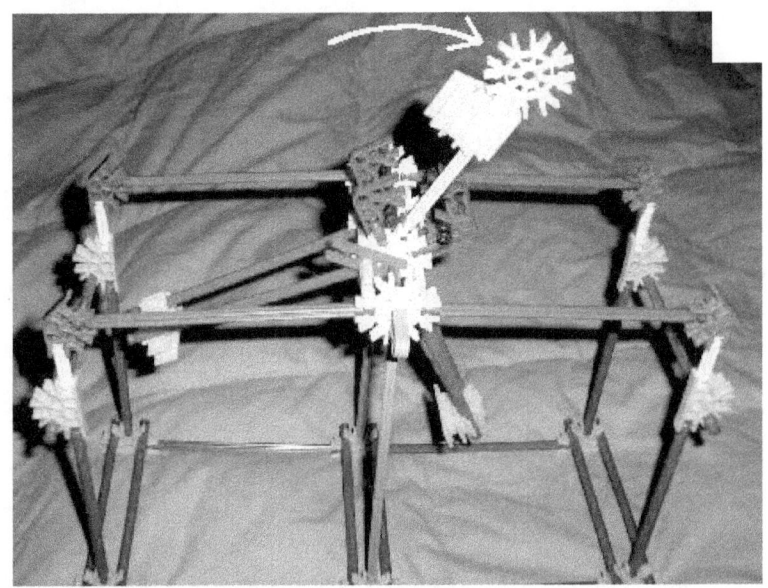

PRINCIPLED ASYMMETRY IN MOTION

PRINCIPLED ASYMMETRY 2

SAME OPERATION AS IN THE PREVIOUS
DESIGN, EXCEPT AT THE RISING AND
FALLING POINTS, ADDITIONAL LEVERS
ARE THROWN FORWARDS

THE LEVERS COUNTERWEIGHT ONE AN-
OTHER, CREATING A POINT OF EQUILIB-
RIUM WHICH DIRECTLY PRECEDES THE
POINT AT WHICH THE HEAVY END BE-
GINS TO BE SUPPORTED BY THE AXLE
ALONE. IN THEORY THIS CONTRIBUTES
TO THE MOMENTUM VIA AN EQUI-
LIBRIZING EFFECT, ALTHOUGH OVER-
ALL, THE APPARATUS IS HEAVIER.

NOTABLE IN THIS DEVICE IS THE VERY
SMALL AMOUNT OF RESISTENCE TO
MOVEMENT.

FRONT VIEW

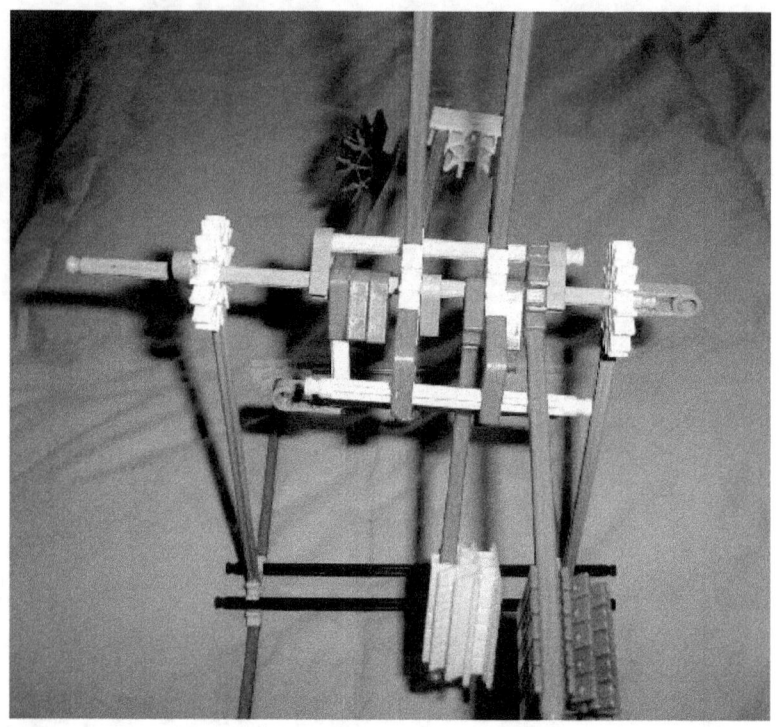

GEARS FROM PRINCIPLED ASYMMETRY 2

SIMPLE TRICKS THAT MAY DEMON-
STRATE PM PRINCIPLES:

1. Gravity and Relation: Adjusting a small oak-
tag sheet between one's hands so that the center weighs on alternating sides allows pushing
the hands to flex the center in alternate directions, without having to move the hands in
relation to the board. Pushing the board in one
direction without return might be compared
loosely to natural limits on motion, since flex-
ing the board in either direction resists return,
if it weren't for the gravitational force.

SIMPLE EXAMPLES: A coin seemingly suspended on edge (above left) shows how common assumptions may not apply to every case.
Here the coin is altered in thickness, resulting
in a case where it CAN in fact stand on its
own, on a level surface. The second photo
(above right) is the case of dominoes which
extend a given movement across space and
through various twists and turns, without in-
creasing the initial force required. Considering
two cases, one in which the first domino
struck a short series, and another in which a
long one is struck, the potential of energy is
extensible independent of initial force. Obvi-
ously it takes greater energy to set up a greater
number of dominoes, but in the abstract it may
be conceived that upright dominoes are sym-
bolic of the potential of mass, whereas a

pushed domino is the active or input principle. Thus extension is more a product of mass-energy than of input-energy (for examples of designs attempting advantage on this principle see the Tilt Motor and Motive Mass Type 2).

2. Slope and Viscocity: in the context of a hoola hoop or a sling, or the center slipping out of a deck of cards, particular methods make enormous differences in effects. This suggests a dynamic whereby special attention of the relationship between slope and viscocity leads to greater energy conservation when compared to the desired result. This has been evident to me particularly when riding a bicycle up a slight incline, where I have found that approaching a slope at an angle is rationally speaking less of a slope, and also that this observation combined with a subtle S-curve in the approach (accounting for balance and speed on the bicycle)results in a near miraculous retention of energy and reduction of frictional resistance.(This is far more difficult at high speeds, in the case of a small ramp). I attribute thisto distributing weight at an angle that is not strictly vertical, in spite of a forward directiveness, such that energy is converted not into the angle of the ramp, but at an angle tangential to the angle, such that the angle on the tire is steep, but in relation to the directivity of the energy, is actually quite shallow. If there are means to using less energy go-

ing up, presumably it is easier than thought to
gain energy going down. For example, I must
refer to an argument I have used in the past,
that throwing energy into a downward slope is
more productive than throwing it into an upward slope. The same is the case with downward trajectories, even if in ballistics this is
not an efficient use of distance. This viscous
effect may also be demonstrated in a tureen or
basin in which water is cycled in a centrifugal
manner.

3. Constructs: Houses use energy to remain
upright; perhaps the same constant energy
might be used towards movement for the same
duration; some say inputting for movement is
different from input for simple mass, in part
because built structures rely on inherent properties of materials; however, one might argue
that a windmill that is power-viable takes no
more energy to support than a similar structure that doesn't produce; thus, where movement is inherent in structure, this suggests energy inherent in mass. That perpetual motion
would follow from this principle is less a mat-
ter of the energy-viability of mass than of the
mobile properties of structure.

Certain Configurations of Blocks
Appear to Defy Gravity

None of these involve glue, although sometimes the lower
block is positioned to favor the counter-balanced end.

[1]

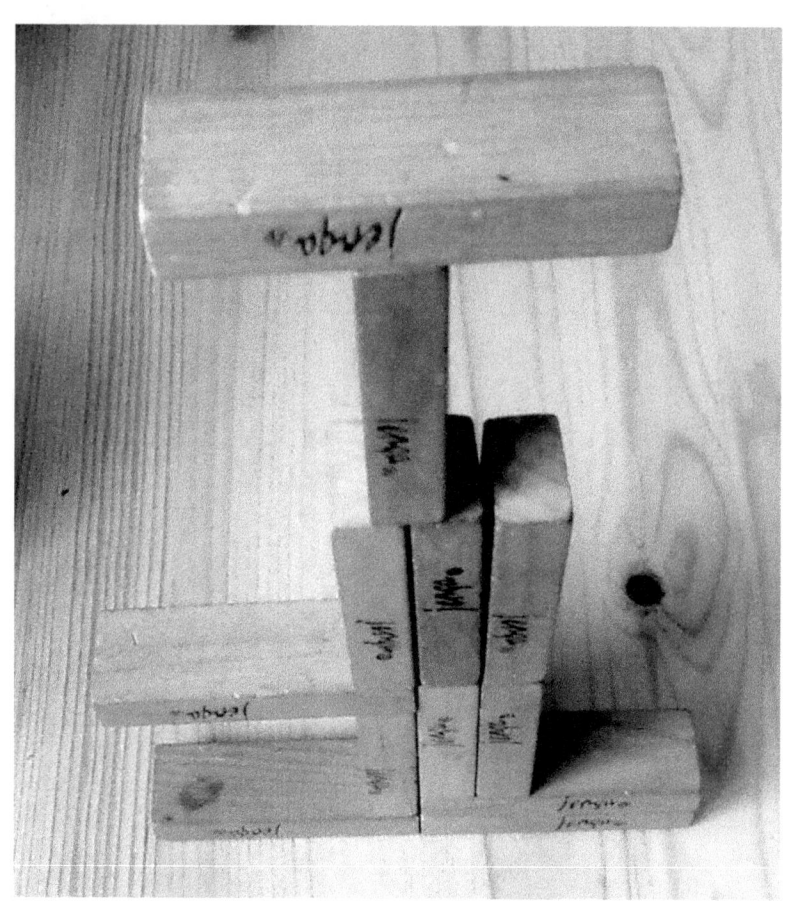

[2]

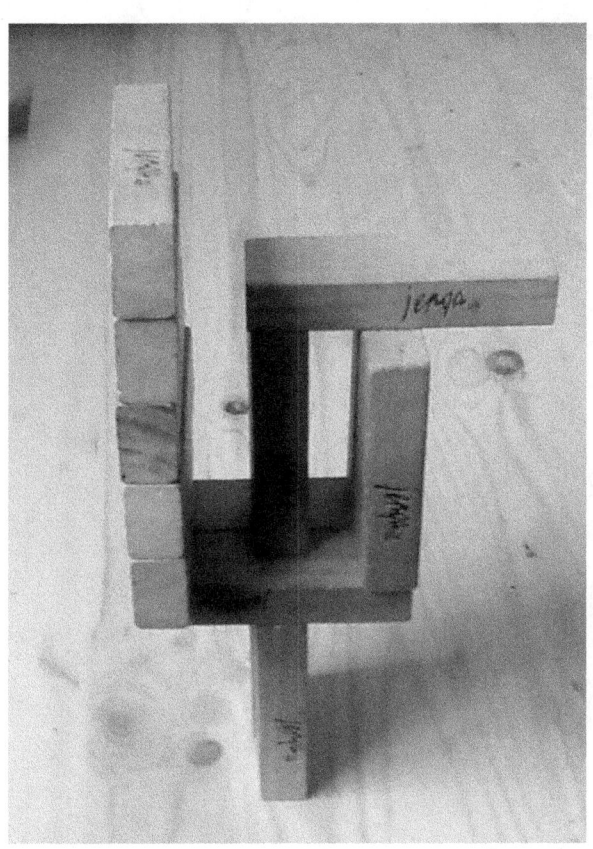

[3]

[4]

[5]

[6]

[7]

[8]

[9]

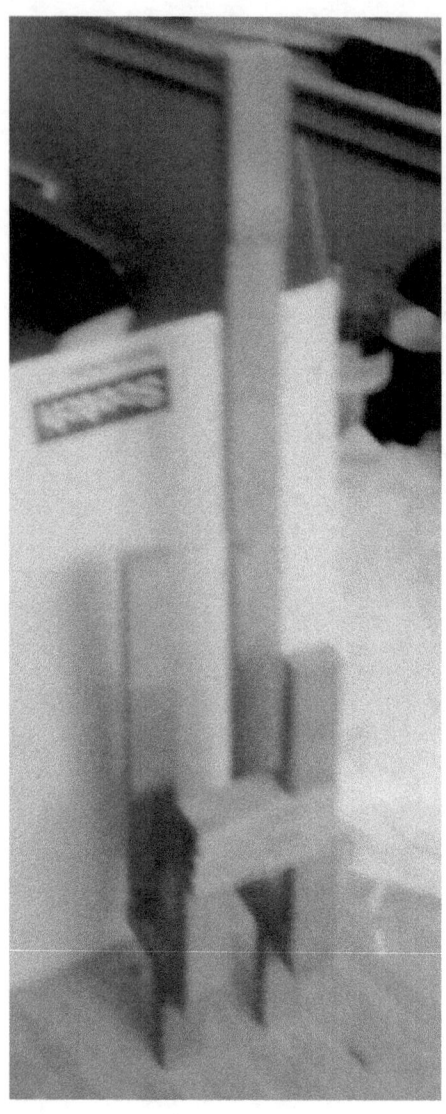

[10]

WERE THE BLOCK CONSTRUCTIONS PSYCHIC EFFECTS?

I have photographed a few other phenomena which might be psychic:

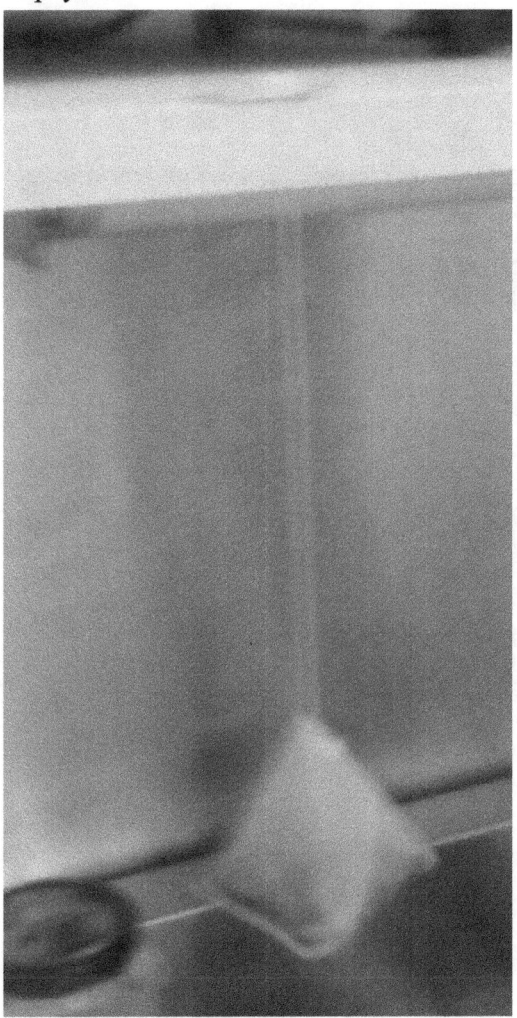

This teabag is really suspended only by moisture on the surface of the shelf! It's a heavier than average teabag! How does it do it?

In this case the string of the teabag somehow caught on the edge of the cup... Weird. It looks like it's suspended in space. And it's supporting the tag on the end!

I thought I had photographed my astral body, until my dad
pointed out that the sleeve is invisible, too!
Maybe my astral body propagates inside the sleeve?

TILT MOTOR EXPERIMENT

A. A wheel with a hand-axle is designed to roll along two planks of art board.
B. As the wheel rolls, the levers underneath each of the boards raise the height of the boards behind the motion of the rolling wheel.
C. The object is to lose no height, while also permitting a rolling motion automatically.

Notation: This experiment confirmed, while later experiments denied the possibility. It is comforting to think that this experiment was more rigorous and well-designed than the later ones.

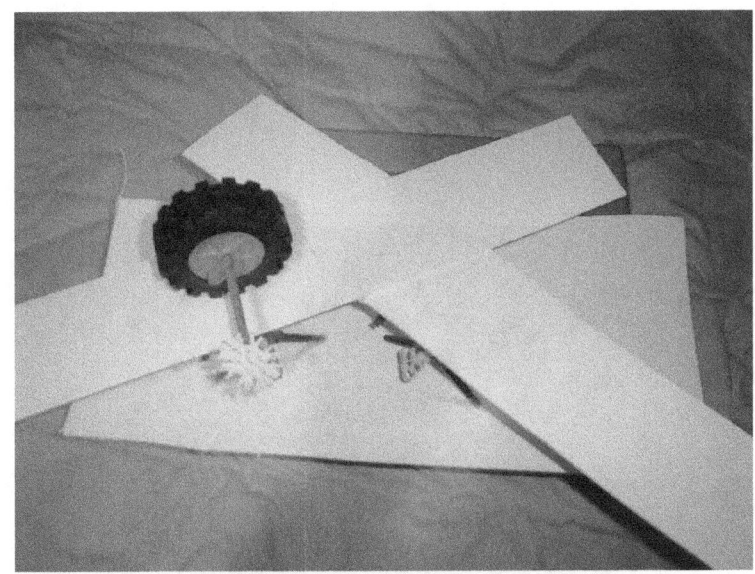

ABOVE: OVERALL VIEW

SIDEVIEW

AFTER MOVEMENT

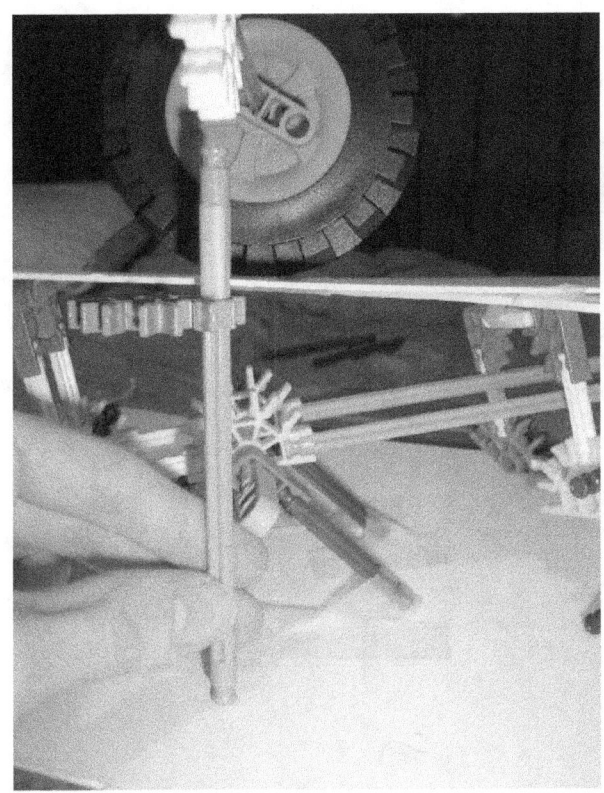

START MEASURE

END MEASURE

EXPERIMENT WITH TROUGH LEVER-AGE

A. My experiment proved that a weight can be lifted vertically through the motion of a member through a slotted track.

B. In this case, the track was moved with a fixed central member due to difficulties of construction.

Images follow:

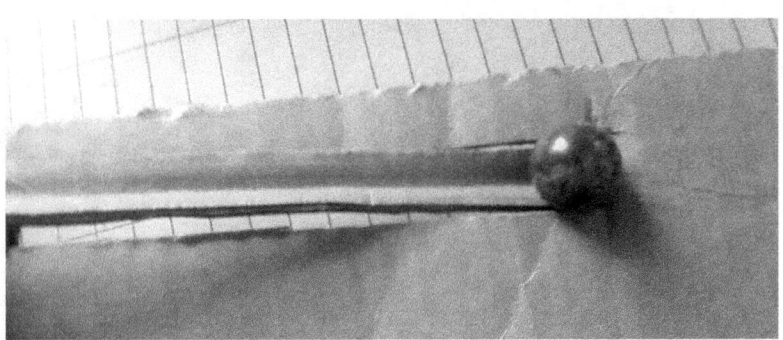

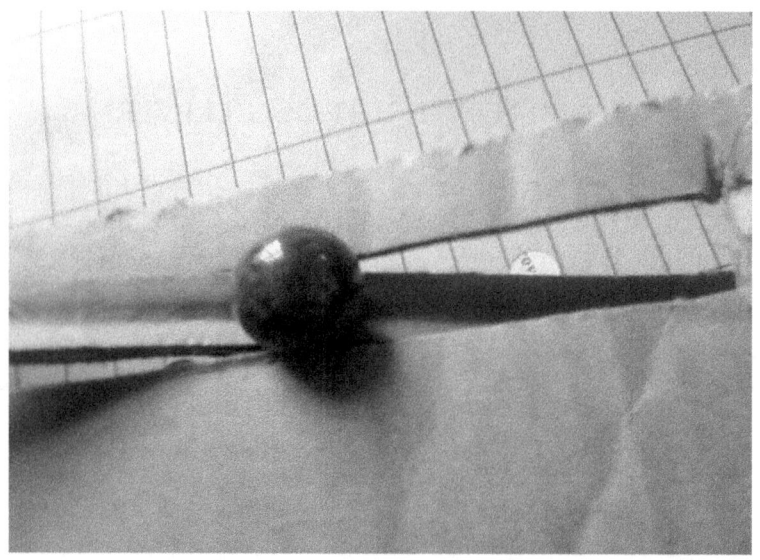

MOTION THROUGH AN IMPLIED COUNTERWEIGHT

SIX PRINCIPLES OF OVER-UNITY (MODULAR TROUGH LEVERAGE DE-VICE)

(1) It begins from rest and uses no electricity or stored energy, except a counterweight, (2) It moves upwards and then downwards on its own, (3) It uses a principle of weight versus leverage, with the weight at a lesser leverage distance, (4) It makes use of a supporting track, which creates in imbalance between the mobile weight and the counterweight, (5) The lever is unbalanced at every point of motion, and (6) All parts may return to their initial altitudes after motion.

SUCCESSFUL OVER-UNITY EXPERI-MENT 1

On November 10, 2013 I experimented with an apparatus I had built designed to test the principle of whether a marble could effectively lift its own weight.

This was done by arranging a counter-weighted lever which passes through a track designed to partially support the marble during its upward movement.

This was a kind of 'cheating' I acknowl-edged, but it was not exactly cheating physics. After all, it is impossible to cheat real, immutable laws. There was only some question as to whether the laws might be different than some peo-ple assumed...

The device proved that an object could op-erate a counterweight after being lifted some distance by the same counterweight. This was a principle that had previously re-mained unproved.

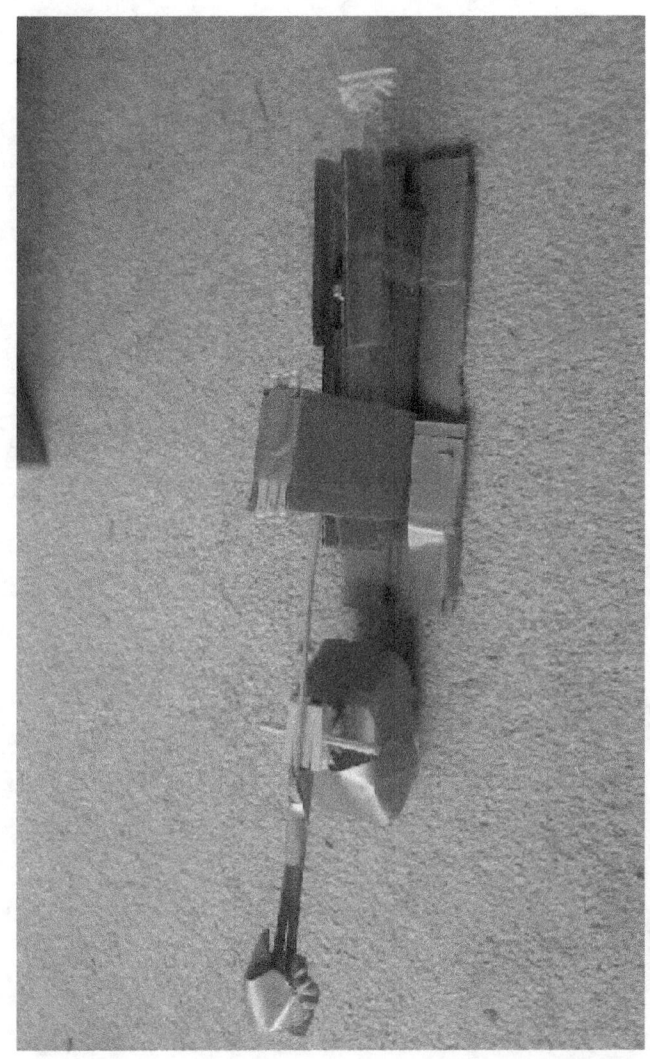

SIDE VIEW OF THE SUCCESSFUL OVER-UNITY EXPERIMENT INVOLVING A TROUGH LEVER

START AND END POSITIONS

A TOP VIEW OF THE APPARATUS DE-
SIGNED TO PROVE OVER-UNITY

MASTER ANGLE / REVERSE GRAVITY 1

"We know that objects in some instances can roll up" ---Aristotle

Pictured: A Master Angle.

I believe I have discovered a specific angle which allows a marble to roll upwards.

In order to defend the very existence of the claim, I will illustrate a very simple argument, and then follow it up with a description of the exact principles.

A rhetorical argument proving the method:

Consider the general case of a ramp.

A sloped ramp is less sloped at a sideways angle to the slope.

For example, in the following drawing,

'A' is less sloped than 'B'. This is undeniable. ('A' has a longer path, but path is not automatically important in every case).

That the angles are different, and that 'A' is less sloped than 'B' can be proven by the angle that is 90 degrees sideways, which is perfectly horizontal. Clearly there are degrees, however, between the two, proving that points which are intermediate have less angularity.

Now we'll look at another diagram:

We know the surface of the ramp is two-dimensional. This is undeniable. So there are

two directions the marble can move: horizontally, and vertically.

However, if the ramp is positioned differently,
with an upwards slope, then it appears there is
a case where the horizontal can be sharper, but
not along the vertical. This is an interesting
property, because in theory it could be used to
develop a perpetual motion machine:

**h. is sharper than v., but
not on   the v.**

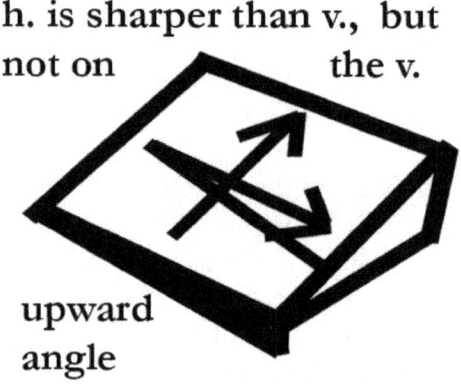

**upward
angle**

Above: here what is shown is the mostly hori-
zontally-sloped support running diagonally
below the approximate horizontal line. However, the ramp itself is also sloped upwards.
When the horizontal slope is sharper than the
vertical slope, but less sloped along the vertical, then any motion introduced does not encounter gravity resistance. Momentum is provided by the angularity of the backboard pushing sideways on the marble.

If you want a more detailed explanation, we
have to look at the specific properties of the
marble:

A spherical marble has less support in any direction when pushed. This is one reason it can be granted momentum without being in the air.

Provided the ratio between the angle of the marble's surface and

(the difference between the horizontally-supporting angularity of the ramp beneath the marble and the angularity of the primary vertical slope)

is > 1,

then sometimes the vertical slope can be upwards when the horizontal is supporting and downwards (horizontally, that is, sideways), as shown in my video.

In other words, I believe I have evidence that a marble can roll upwards, in a very specific configuration.

FIXED PENDULUM LEVER

This model was created in November 2015

Experimentation seemed to show that a marble could be lifted through the operation of a sideways-angled pendulum that is fixed within a particular range of motion.

SPIRAL PENDULUM

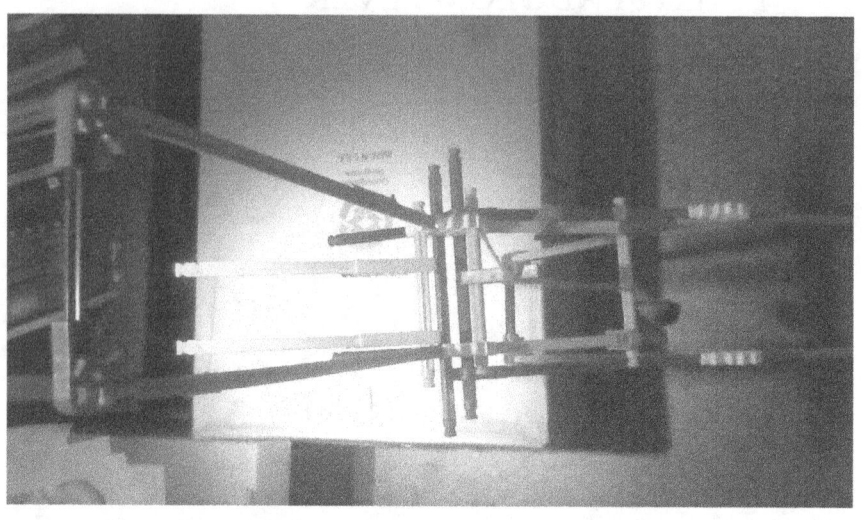

TOP VIEW

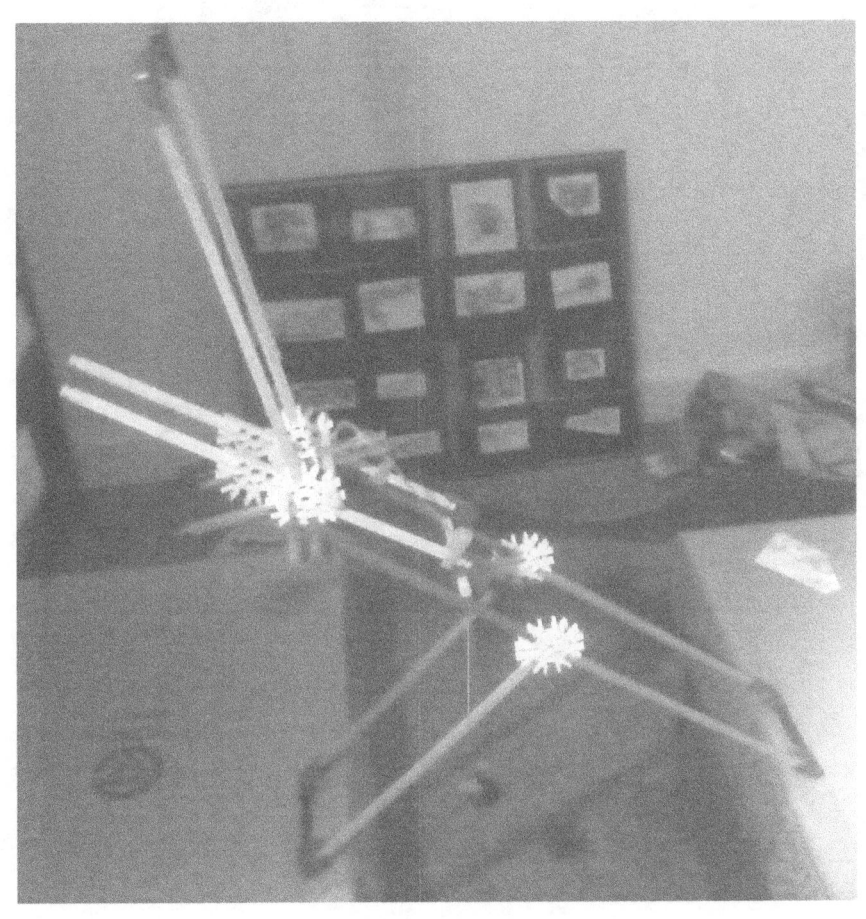

FRONT VIEW OF THE SPIRAL
PENDULUM

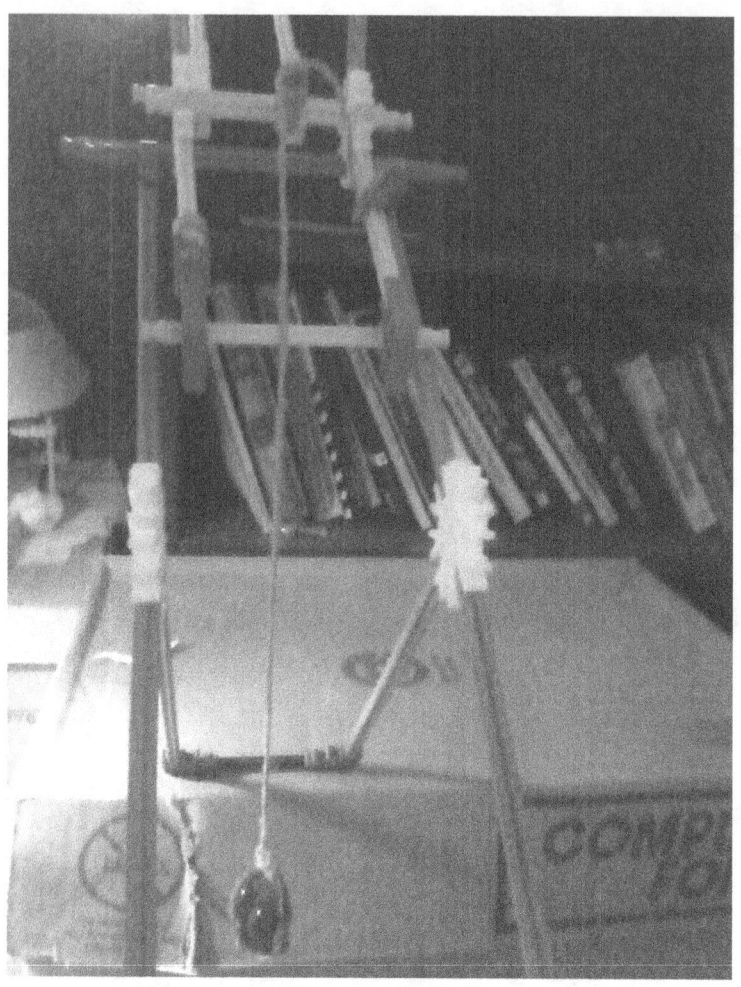

SIDE VIEW OF THE
SPIRAL PENDULUM

An ordinary pendulum is said to swing a maximum of 168 times. This pendulum swung 225 times. It is designed using a spiral for efficiency, and stops inhibiting excessive movement.

OVER-UNITY PENDULUM (A DIFFER-
NET DESIGN)

This offset lever pendulum spun 360
degrees after a 180 degree hard-turn,
which is not commonly known to
happen in physics!

PRINCIPLES OF SPECIPHYSICS

[Based on the Master Angle concept]:

1st Law:

"Where principles interact in multiple dimensions, and the principles are not opposed in motion, at least one common principle remains between the two, which is best abbreviated as neither of the two prior principles in ideal cases"

2nd Law:

"Where the properties of an object are multiple, a common principle can have multiple properties of such an object"

3rd Law:

"Where potential permits (e.g. where motion in space or time is permitted), the only limit is mechanics, not entropy. If entropy is proposed as an explanation, it can be said to defer to this law, which is more universal"

end

POSTSCRIPT

I HOPE YOU HAVE ENJOYED MY NU-
MEROUS DESIGNS FOR OVER-
UNITY MACHINES, AND THE EX-
CITEMENT AND ENGINUITY THAT
WENT INTO FORMULATING THESE
MANY POTENTIAL EXAMPLES.

IN THE FUTURE PERHAPS SOME-
THING EVEN MORE EXCITING WILL
HAPPEN!

AS FOR NOW, WE HAVE TO LOOK TO
THE MOTIVATIONS OF NEW HOBBY-
ISTS AND INVENTORS TO REALIZE
THE DREAM OF OVER-UNITY!

NON-FICTION BY NATHAN COPPEDGE
How to Write Aphorisms
Nathan Coppedge's Perpetual Motion Machine Designs & Theory

FORTHCOMING:
"SCIENTIFIC THEORIES"

SERIES BY NATHAN COPPEDGE

The Dimensional Encyclopedia
The Perpetual Motion Genius' Guides

PHILOSOPHY BY NATHAN COPPEDGE

The Dimensional Philosopher's Toolkit
The Ninesquare Notebook
Intermediate Insights
Modal Dimensionism
Arche-Logos: ACEPM

FICTION BY NATHAN COPPEDGE

The Lessons of the Master, by Master Kuo
The Story of Master Wu, by Master Kuo

BIO

Nathan Coppedge has been quoted in Book Forum and the Hartford Courant, and is a "famous quotable" at Poemhunter.com. He is a member of the International Honor Society for Philosophy. He has run a website on perpetual motion machines since 2006.

www.ingramcontent.com/pod-product-compliance
Lightning Source LLC
Chambersburg PA
CBHW071629170526
45166CB00003B/1250